ROSHAN CIPRIANI

Living The Mandela Effect:

Worldwide Reality Shifting

A Guide To Staying Sane And Surviving Your New Reality

Copyright 2016 Roshan Cipriani
The spiritual and moral rights of the author have been asserted.
All Rights Reserved
No part of this book maybe used or reproduced by any means
graphic electronic or mechanical including photocopying, recording taping
or by any information storage retrieval system without the
written permission of the author.
ISNB-13:978-1537175911
ISBN-10:1537175912

Table Of Contents

Dedication ~ *To Jade*

Preface ~

Introduction~ *Dimensional Travelers*

Chapter One ~ *Thought Energy*

Chapter Two ~ *It's Real Somewhere*

Chapter Three ~ *The Soul And Its Confusion*

Chapter Four ~ *Technology Gone Wild*

Chapter Five ~ *What Do We Do Now*

Dedication

This book is dedicated to my magnificent and wise daughter, Jade.

We have traveled many earths, galaxies and dimensions together, often seeing many strange and peculiar things.

I know we will continue to hold each other's hands as we jump to many more.

Be safe always.

I love you.

Mum.

Preface

This book is a simple *summary* of extremely detailed research and evidence. I found it was necessary to simplify and dissect some of the answers to the questions in an attempt to present the information.

There are many repeated statements throughout this book due to the nature of the inquiry, and I am grateful to be able to access the same information with the answers to multiple questions.

In reading and seeing the truth on this subject there may be times you feel the truth is more frightening than any fiction you have ever heard. I believe that there is no overstatement and any repeated emphasis can be appreciated and valued on so important and monumental a subject. I want you to know that you are experiencing something much bigger than yourself and what is on the line here is much too important to be ruined by insensitive people or difficult situations. Unfortunately, you may encounter ignorant people and you must have the fortitude to keep on seeking and learning.

There is no 'perfect' way tackle this topic. We all vary in how to approach it, usually influenced by past and present experiences plus our faith/ politics/ personal values.

There is only one way out of this dilemma, and that is to move beyond the comfortable illusion that we can continue on without telling the truth. Speak your truth as you know it no matter what. To those who hold to a high standard of religious observance, remember that *true spirituality covers all of reality.*

Roshan Cipriani, 2016

Introduction ~ *Dimensional Travelers*

Quantum physics teaches us that our physical reality is moldable, and that our physical universe is shapeable. Quantum physics research proves that particles of light and matter behave differently when different human beings observe them; no two people can observe them in the same way. Quantum physics has proven that all matter is energy.

It is nothing more than moving energy. Quantum physics has proven that all energy oscillates between being incarnate and not being.

The laws of quantum physics reveal that everything must be formulated first in the interior sphere. The internal materiality is the frame, the diagram, or divine map, and then physical matter is enticed to this "configuration" by the laws of quantum physics. Now we understand that approximately eighty –five percent of everything in creation is thought waves.

These thought waves I believe are the thought waves of the Most High God. Although some people may say that there is no proof of His existence, there are myriad documentary evidences throughout every sphere of scientific inquiry, which clearly demonstrates the opposite to those who look for them. Those truths are hard for the natural mind to receive, but Scripture is unequivocal. The Most High God controls all things.

The argument against the concept of divine sovereignty is an old one. *"You will say to me then, 'Why does He still find fault? For who resists His will?"*

In other words, doesn't this sovereignty cancel out all human responsibility?

The scripture itself has the answer-

"On the contrary, who are you, O man, who answers back to God? The thing molded will not say to the molder, 'Why did you make me like this,' will it? Or does not the potter have a right over the clay, to make from the same lump one vessel for honorable use, and another for common use?"

In Psalm 50:21 the Most High G-d says, *"You thought that I was just like you."*

However the Most High God is not like us, neither can He be understood to our standards.

Without any pretense of comprehensiveness, this book will try to depict my own as well as other travelers' experiences and observations of *Living The Mandela Effect: Worldwide Reality Shifting* and explore what may have caused all of this to happen at this time. I sincerely believe that you will benefit from the information within these pages.

There is a gap that we must cross between our old life and our new life which involves surrendering our old efforts and way of existing in this world. If we don't our procrastination just delays our acceptances of the inevitable changes that now engulf the entire world and makes living harder for us.

Definitions are often a matter of life and death. So it is possible for two people to use different definitions of a term, without disagreeing in their legitimate views. I will however try to use the most widely and recognized definitions, to maximize comprehension.

The Mandela Effect is a theory of parallel universes, constructed within the idea that many people have similar alternative memories about past events, and they may all have been in a different dimension with different occurrences and outcomes and not be mistaken in their recollections. The Mandela Effect was first designated as such online in 2008/2010. A woman named Fiona Broome discovered that others had a distinct and different memory from the norm, and it was similar to hers. She along with many other people remembered that Nelson Mandela had died during his imprisonment in the 1980s and was confused as to how he could have now died again in 2013. This revelation opened the conversation and people began revealing other strange recollections not always shared by people around them, yet something they have in common with many other people in the world.

Having experienced other alternate memories and realizing that I wasn't alone – and that many other people have shared the same alternate memories was just the starting point in my research. However I believe that someone has altered something that has impacted our universe and apparently other dimensions as well.

This book should be taken as an outline, a first stride, towards a completely different style of awareness and thinking in order to handle the traumas of these extraordinary experiences. The science supports the possibility that this could and has happened.

This is because living things have been tied to the Earth's natural magnetic field from the time that life began. This magnetic field permeates and contributes to all life on the planet by the generation of what we call the atmosphere. This field varies in strength and consistency through the ages and, with this variance, so too does life on the planet change. Our bodies own magnetic frequencies and bio-field patterns react to this variance of the Earth's field, so we can either learn to adjust our reactions and go with the flow, or continue to push back against our new experiences to our own harm.

Since we understand that all matter on Earth assists in creating this field and so becomes charged with this magnetic resonance; displaying different results we know that we too can display different results depending on our own magnetic resonance or field charge.

As all matter is made up of magnetic resonant field patterns, of varying strength and frequencies, applying a magnetic device of any configuration will produce an effect one way or another. Today humanity's use of electromagnetism for power and communications has produced an abnormal electromagnetic environment unlike anything that has existed before.

Everyone is exposed to a complex mix of weak electric and magnetic fields, both at home and at work, from the generation and transmission of electricity, domestic appliances and industrial equipment, to telecommunications and broadcasting. When electric fields act on conductive materials, they influence the distribution of electric charges at their surface. They cause current to flow through the body to the ground. Electromagnetic radiation causes tissue damage by releasing electrons in the cells, called ionization. This is done by radioactive particles or electromagnetic waves with sufficient energy colliding with many electrons on the atom to knock electrons off the atom. The electron expelled off the atom is called the primary electron.

When the primary electrons hold energy, a particle ejecting the primary electron may cause it to eject another electron, either on their own atom or on another atom. This is known as secondary ionization.

These are the forces that are contributing to the strange effects felt and seen by those of us now seeking answers.

Due to the electromagnetic wave spectrum and its effect, divided into ionizing radiation such as ultraviolet and X-rays and non-ionizing radiation such as radiofrequency (RF); Wi-Fi, cell phones, and Smart Meter wireless communication; ionizing radiation breaks the electron bonds that hold molecules like DNA together.

In other words it can interfere with the very essence of who you are at the mind and cellular level.

Since brain cells receive nutrients and energy from blood and body fluids, any change in the blood and body fluid atmosphere concerned with capacity, pressure, temperature, or chemical responses will pose potential harm, or death, to brain cells.

The normal blood volume of a body is about five liters, and the brain requires a continuous supply of glucose from this volume to give energy to its nerve cells, however a decrease or an increase *can create unprecedented changes in ways we are now beginning to comprehend.*

Since the brain will react to each and every external stimulus sensation, the now sensitive tolerances will become even more sensitive.

Nikola Tesla once said: *"If you wish to understand the secrets of the Universe, think of energy, frequency, and vibration."*

The ladder-like structures that make up the double helix of your DNA represent the chemicals that tie the two strands together. These arrangements contain the genetic code, and the sequence of the bases communicates what kind of molecule is required for building the protein that is needed. Scientists found that an electromagnetic field applied to lab rats over time resulted in broken strands of DNA in the brain. Imagine that field applied twenty-four hours/seven days a week to the human body and you can understand the damage being done to the human DNA that could have made us susceptible.

Electromagnetic pollution is caused by electromagnetic field or EMF radiation. Today, we live in a world full of electromagnetic pollution problems.

The geomagnetic field is a shield protecting the Earth from the full force of the sun's energy. Without it, life could not exist. But since humanity has created other forces beneath this shield now we understand that the electromagnetic fields also affect us in significant ways. This force has the ability to impact the soul, mind, consciousness and will of us human beings. It may also have the ability to somehow transport us back and forth across simultaneously existing versions of our many dimensional lives, or anchor us firmly into one particular life.

We know that thoughts, feelings and consciousness are seen as a form of electrical activity in the brain and can be measured using an encephalogram; so this energy is not something theoretical and unverified. Yet the mechanics may not really be fully understood.

Positive ions in the atmosphere have harmful effects on health, whilst negative ions are beneficial. When our cells are in balance, they are predominantly negatively-charged, and extensively use elements such as calcium, potassium and sodium ions for keeping our cellular structure in place. These chemical elements are used to transmit electrical signals between the brain and the nerves so that our bodies function as needed.

Today modern man has filled up the electromagnetic spectrum from visible light to zero frequency, which had remained empty over the eons of time. We know that our bodies have not adapted to this, by the increasing waves of unexplained complaints and discomforts that have accompanied the constant bombardment.

Then there is the audible spectrum. Sound waves are also known as sound radiation. We use ultrasound in medicine to take pictures of soft tissues. Ultrasound means above the sound levels we normally hear. Computer monitors produce ultrasound, and some very sensitive people actually can hear a high pitched sound coming from their computers, and it is not an isolated event. Yet could it be of greater importance than previously thought.

Modern physics has two basic scientific laws: quantum physics and general relativity. These two scientific laws represent radically different fields of study. Quantum physics studies the very smallest objects in nature, while relativity tends to study nature on the scale of planets, galaxies, and the universe as a whole.

In the most basic quantum physics, the non-charged electrons are in a para or normal state. When magnetically charged, the excited electrons of the atoms change their orbit to a spin parallel with the pole surface and they take on an energized condition known as the Orto state. In this state, the nucleus changes direction against the spinning electron creating a friction that is measured in an increased or endothermic (transport of energy) electrical value.

Powerful permanent magnets have their matrix altered by the artificial energizing processes used, and do not conform to the natural complex magnetic field structure of the human bio-organism, or that of the planet. Now with sites like CERN and the huge Magnetic structure at the LHC in use continuously it has only become more critical.

These processes use artificial electrical energy, which has within its structure a frequency composition different to that of nature, and nature will demonstrate this by being altered.

This is what I believe has happened for many of us who have realized and experienced this "shifting effect." There are massive disturbances in the magnetosphere of the earth when the (LHC) Hadron Collider is turned on. Something unexplainable has happened, and the effects are far reaching into areas no one has yet considered.

For numerous people the world seems to have changed. Perhaps when a dimension/ reality becomes discordant with your continued existence; you die there, then your consciousness "merges" to the adjacent useable dimension/reality in which you are still alive, and you continue to live on as you -- but with subtle differences. A new side effect of this process is that the new dimension/reality is now actually inconsistent with your memories. This is because it was completely different to your memories of the one you left. Previously old memories were not remembered at all.

When this happens repeatedly and when you find yourself basically a "traveler" within multiple versions of "your" life, which you now remember.

Sooner or later you'll have to come up with a plan on just how to handle the strain and madness of what is at present your "new" life. The way I see it, we seem to be able to shift in between different dimensions/realities and they don't seem like they are "correct" or "normal" somehow to our own remembered experiences and we often don't know what to say or do.

If this phenomenon lasts or increases a lot more people are going to find themselves in a world they no longer are familiar with asking questions that no one there can answer.

For instance, what if you wake up one day and the country you lived in no longer exists but you have clear memories of being there your entire life? What if your spouse or parents don't even remember there was such a place? Do you just accept their memories as more valid than your own? Do you just forget what you KNOW since you cannot explain it even to yourself? Could you just decide it must be your "faulty memory" if in the morning at the breakfast table one of your children or even your present spouse just doesn't exist? Do you ask about the person only to be greeted with blank stares and then denial? After you get over the shock that a firm memory you had just doesn't exist in this world, would you then try to figure out what has happened? Would you wonder if something is wrong with you?

The idea of abandoning what you thought you remembered in favor of what other people state as true in order not to be thought of as crazy can be actually quite frightening.

That idea is actually becoming a daily reality for many people. Additionally it takes a large leap forward in thinking to consider that the fabric of reality shifted at some point in the past, and we are in some type of parallel inhabitable reality/dimension/ universe, but that is preferable often to the unacceptable alternatives. It even seems often that we are continually moving back and forth between dimension and realities, as we see things altered and then watch as they return to their original state.

The only proof we have -- if ***WITHOUT A DOUBT WE ARE IN A NEW REALITY/DIMENSION/UNIVERSE,*** are the **_memories_** of those of us who for whatever reason shifted to this reality/earth/dimension/universe, and kept those old memories. There may be multiple contributors to this but I believe that the LHC at CERN in tandem with whatever D Wave quantum computers they have searching the universe starts blending with **every** other dimension that has a CERN and an LHC **doing the exact same thing at the exact same time,** is the primary cause.

It would all have to actually synchronize for greatest effect. These are similar realities/dimensions/earths yet they are not the same.

The dominant frequency and the place on the earth where the shadow of that dimension overlaps with your current reality perhaps, when the power is released, determines what set of people from our original reality are relocated dimensionally and exactly which reality and dimension they are sent into. They become "CERN /LHC Refugees" and "Dimensional Travelers," without even being aware of what has actually transpired, just feeling things are somehow not right.

We must stay strong and be brave in the face of this new life, which can and will change in an instant. Finding and listening others experiencing the same thing is just one way to do this.

One cannot prove nor disprove the Mandela Effect. The whole principle behind the phenomenon is that our reality has one way or another changed. It doesn't matter how far back in time you look, the present reality will reflect what was true for its dimension/reality without any alteration that would support the different memories. If you are living the Mandela Effect you already know this.

You are somewhere else, and in this new place this is the way they remember it here. If you don't see any changes or differences and everything is just as it always was, it's because "you're still in Mayberry, Aunt Bee."

Don't expect understanding from people for whom everything is the same, and you won't be disappointed or hurt by their reactions.

Often unless a person starts having similar experiences, you will just be considered odd, mistaken or even obsessed with speaking about "that crazy stuff."

If there are many other versions of reality/dimensions and universes, could they also have slightly different physical and scientific laws as well? What about their spiritual laws? Could they be different as well?

You can realize that this is a very serious phenomenon, when it impacts **EVERYTHING** about your entire life and existence.

What do I remember about my original home? I remember happy times, good and kind people, praying and family worship, driving along the coast, clean air, clear blue skies, a bright yellow sun and so much more. Though I have been transported far away from the familiar, my memories are still strong and sure. I've been tense and I've been heartbroken over much of the reactions I have seen others like myself receive and I miss my original home so much. Yet in the face of so much confusion I continue to just go on hoping to get answers each day.

I do know one thing for certain; as this phenomenon seems to be expanding, there will be many more people worldwide who suddenly will begin to understand what this is all about on a personal level when it hits home for them.

Chapter One

Thought Energy

Even when you know how something will eventually happen you still have that sense of anxiety and apprehension. You wait nervously for it to happen. Then when it begins you just cannot believe it is actually occurring right in front of your eyes. The longed for moment, the awaited event is here. Yet often there is some component that you never considered, and may have overlooked. This is where imagining comes in, which is merely a method for generating a mental duplicate of a forthcoming experience. When we visualize our anticipated outcome, we begin to "see" the chances of attaining it.

You may have picked up this book because you've gotten to a juncture in your life where you're desperate for more than just the customary theories and explanations about what is happening and you want to understand more. Even a child has the capacity to understand when something is wrong and to attempt to get away from hurtful and uncomfortable circumstances, and yet much of humanity is seemingly bent on hurtling rapidly to their own pain and destruction.

There is also an awakening happening all around the world. And that awakening involves you and every being that exists on the earth. Our wisdom stems from coming to understand how little we theoretically know about life, the universe, the Most High G-d and everything else that affects us.

So I am encouraging you to envision mentally a positive outcome to all that you are experiencing right now. This Universe is Intelligent; not in an academic sense but in a way of manipulating substance / matter for the benefit of all it contains and surprisingly the benefit of other less fortunate Universes and dimensions.

Quantum physics demonstrates that we live in a space-time continuum in which all events occur simultaneously. So at the time you are asleep in one dimension you may be in deep conversation with an ancestor, or a friend from elsewhere at the same time. Spiritual teachings also tell us that we have a reality beyond that of this physical world.

"Deja vu, (together with *Deja vecu* and *Deja Visite*,) among others is a French term meaning "already seen," and is considered an interruption between objective unfamiliarity and familiarity. I believe that *deja vu* is a quantum phenomenon. With *Deja vecu*, you may even feel that you already know what's going to happen next. It is possible that *Deja vus* are "memories" from other versions of us in other universes, or mainly memories from different dimensional timelines. Signs are anything that "stands out from the norm". Anything that catches your attention and there is no reason why it has happened is a sign.

 Seek out the truth; go with your inner feelings. You must pay attention to your thoughts and to the world around you. Just because you can't see the connection or reason for something happening does not mean there is no connection.

There is always a connection. The more you pay attention the more you will see. Synchronicity is a noteworthy, highly improbable simultaneous event that is revealed to you.

You are not alone in this, because we are all connected in ways we don't understand, and sometimes an area of that connection may resonate enough for us to take notice.

How do I know this? Basically, I am forced to put this idea to the test daily.

The first and foremost principle of Quantum Physics is that the observer affects the outcome every time. Your mind and the subconscious awareness create in your life. Everything that you can see in the physical world is made up of molecules that you can influence. Thought energy is the energy created by ideas, feelings, intentions and beliefs. Einstein acknowledged over eighty years ago that matter and energy are interchangeable. Yet, many of us do not understand what that really means. Thoughts alter into subatomic particles which are directly affected and take physical shape constructed on the thinking and belief of the person thinking them. The energy of thoughts can even be measured using modern devices such as PET scans (Positron Emission Tomography) and MRI (Magnetic Resonance Imaging) so we know it is quite real.

You might be surprised as to how effects like hyperspace, physical portals, and multi-dimensional existences can be affecting our daily lives and our thinking.

There are many of us who are convinced that what we remember is real and we witnessed things that now we are told just never happened. When two dimensions collapse, or collide together, the dominant dimension seems to take over. Perhaps time streams were not just separated by time, but by particles of matter and these are the differences in physics itself.

Technically speaking, there would have been an infinite number of progressive dimensions, each individually divided by little alterations. But due to the fracturing of the time streams; events and things have changed and people are in strange and unfamiliar dimensions with strangers that look like the people they knew, and situations that are vastly different than their previous reality and recollections.

It seems that the people from a different dimension or reality/time stream for some reason brought their memories intact with them. So if you remember everything exactly as it is now you would then be in the dimension or reality/time stream in which you have always existed, so everything is as it has always been for you.

There are many parallel universes and perhaps they could in theory collide with each other and become somehow fractured, resulting in the transference of some type of magnetic energy and with it some of the people from one dimension/ reality into another without anyone realizing at first what might have occurred.

I know that the word "multiverse" is used instead of "universe" because scientists have accepted through quantum mechanics and the string theory that there truly are more than one dimension and more than one universe.

The idea that Einstein is actually right; and that when you match the frequency of the reality/dimension you want you go into, then that reality/dimension becomes your new reality/dimension is really quite astounding when it may be happening to you. Today modern scientists have ignored the fact that when something changes, it changes not just here or there, but EVERYWHERE. You could then enter a completely new time stream of consciousness where the whole past is different. Consequently when you look back into the past, it too has completely changed.

That's why there are no newspaper clippings, paintings, advertising commercials, and even your own DVD movie collection that you can access to verify your own memories. It has all changed when you did and the possibility exists that you may have done more than physically move. Any two realities that share a timeline close enough can stream-slip together. Anything that is the same between the two will merge. There is so far no way to say how that is determined.

So if you died in one reality, undoubtedly the reality dimension/time stream on which you are still alive has taken dominance. So for anyone who's new to this idea — and slightly astounded, identifying many things as different now — yes, it's okay to feel unstable.

There is a well-known disaster theory which pictures a precarious universe, and the Mandela Effect may represent even bigger dangers, a perilous multiverse of dimensions and time lines no longer isolated from each other.

We are literally submerged in an ocean of electro-magnetic radiation. Is it possible that we have somehow inadvertently technologically interfered with the dimensional frequencies and time streams of other planes of existence? Could every dimension/time stream consist of its own distinct frequency?

The possibility does exist that when we began to cross the thresholds via massive quantities of subliminal atmospheric pulses of energy into our space, we altered other distinct frequencies. Then there is the theory in which many people feel that they are not in their respective original "home" dimension /time stream, after they had a life-threatening occurrence where they might have actually died.

Within this theory, it is further suggested that when the disaster was inescapable, people were somehow taken from that original "home" reality/dimension time stream and placed in the adjacent "comparable home" reality dimension time stream. Often it is the one they find themselves living in quite uncomfortably, and constantly questioning.

Could there have been an ELE (Extinction Level Event)?

If we take one interpretation of quantum studies accurately, every possibility creates a new universe dimension/reality.

So, in at least one reality, people are fully aware of their having been an ELE, and being "removed" to another dimension in which the situation for survival was better.

It would also mean that there is at least one other new universe dimension/reality in which people are being "removed" to another dimension/earth to escape a possible inevitable ELE and they are totally unaware of it and just continue living as usual. They may just remember everything from the old original "home" reality dimension, and only question themselves when finding large groups of others experiencing similar recollections in perfect alignment with their own.

The first law of Thermodynamics specifies that energy such as the electrical charges produced by your brain, or the heat your body puts out - cannot be created or destroyed, but only alters in form—implying that the energy that powers your body must go someplace when you die, and since our consciousness cannot be destroyed, but is infinite you have to continue to "exist."

In the "Many Worlds" explanation, instead of collapsing from superposition into a single reality, the wave function branches into multiple realities comprising each possible outcome. This understanding of quantum mechanics conveys with it the fantastic allegation that all possible pasts exist, each contained within its very own universe. So that every time a decision is made, another complete universe branches off from this one.

The inference is that, among the endlessness of universes being created, we will all follow the precise branch that ensures our immortality. That doesn't mean you never died. In that other universe, your family did mourn your demise. Although in this one you lived as though you just had a close call or NDE (Near Death Experience) and shrugged the danger off as nothing.

Supposing this quantum immortality is valid; there could be multiple physical universes like our own existing side-by-side where consciousness does not end at death. The idea of quantum immortality violates no known laws of physics and quite scientifically is supported by those same laws.

Some say that irrespective of the cause of death, if the many-worlds explanation is true, then there will always be at least one branch/dimension or reality where the "miraculous survival" circumstances has occurred, and that version of "you" will never die. Quantum immortality could mean that you can never start down a road where you end up in a situation with a no chances of survival. This would also mean that by being close to one another, two people would be less likely to die in the other's world by excluding any mishap except one that could only take you both together.

Could there are people who were "moved" by divine involvement with a similarity of purpose, possibly to a parallel universe? It seems that any two realities/dimensions that share a time stream close enough can merge together.

Anything that is the same between the two will amalgamate. These would be the closest in proximity and presumably having diverged from each other previously in their time stream dimensional differences. There may be a merging of parallel universes into one another taking specific people from different dimensions/realities into this one different dimension or reality or even into several others to await a distinct event or sequence of events. Could this be done based on their particular thought patterns, or even their DNA?

I believe the evidence is there for the existence of multiple dimensions and time-streams, multiple harmonic universes that are analogous yet reliant on each other. A reality/time stream dimension shift of some kind is and has occurred. There are many ideas and theories, but one thing is true; if we exist in one reality and we witness changes to this one reality, then something is causing the changes either intentionally or accidentally. It seems we are long past the point of denying that some of us are experiencing altered realities.

Chapter Two

It's Real Somewhere

This experience can really be all about expanding our consciousness. However, we will first have to learn to handle the shock of seeing our reality become something different than we have known.

We are able to realize that something has changed, and we can theorize on what is responsible for the changes to our reality, but ultimately this is something even bigger. Three years ago I began to experience illness and muscular pains, and then I discovered a major freeway overpass, next to my home-- seemingly constructed in one week in my old hometown. There was no way that something like that could have been constructed in one week without me knowing and seeing it. Yet there it was.

I felt sicker with each passing day, and all I wanted to do was sleep. I felt as though my body was incomplete somehow. Then without explanation everything would clear up and I would be completely fine for a few days.

That's when I would go out, and discover that everything in my world seemed off somehow. Then for a short while I had been experiencing an assortment of events involving time. My days either felt slowed to a crawl or would speed by incredibly.

Recently, I have noticed that drives I take and errands I do every day seemed to take less time somehow.

For me it seemed as if time is not stable and the days are shorter, with many weeks passing by in the blink of an eye. Often it's already a new month before I even realize it. This whole thing has been an incredible roller coaster ride. Crazy for me is irrelevant; as personal perception is what matters most to me. I do not make any claims to KNOW what is happening and why, but when something affects your life it becomes important to you. Has the world actually changed, or is it us that have changed and so then also our perception is different.

If the concept of infinity is factual, then what is possible is anything you or anyone else can legitimately imagine, conceive or think of isn't just a possibility, but inevitability.

If you take the predictions of quantum mechanics at face value then everything that can happen does in very precise, a many-worlds kind of way. However that inevitability is actually guaranteed to be occurring somewhere. To my understanding everything that can happen does and the universe has become a pretty bizarre place. Since electrons have the ability to travel anywhere and at any time as per quantum mechanics which states "particles have some small chance of jumping across the universe."

During the course of my life I have witnessed dishes just fall over, pens roll off of tables and other things that seemingly shifted without my interaction with them or altering in some way their environment.

I had never previously considered that they may have moved in my absence and for some reason immediately returned so my consciousness would be able to perceive their existence in my universe. Could these events be a form of Quantum Entanglement/Quantum Teleportation that I observed without having realized what was occurring?

If quantum particles can travel faster than Einstein's speed of light, then ostensibly jumping to a different place in the multiverse and back can happen with no concern about how quickly it can occur. Maybe they turn into waves to travel, as probability waves do not have mass and therefore cannot be restricted by the speed of light and can move much faster than that. Some people will try to account for this unexplained phenomenon by using metaphysical terms or spiritual phraseology to account for something we often cannot begin to comprehend.

I understand that sometimes there are completely ordinary and lucid reasons for what at first appear to be strange and unexplainable events, however for those other times it would appear that we are still left with no new answers. However if a particle in wave form can travel quicker than we could imagine and really be in several places at once; then perhaps we also could move from one place to another traveling the entire Universe instantaneously without being aware of having done exactly that. This could explain why so many of those affected by the Mandela Effect have observed things that seemed changed; seemingly alter back to what they previously knew.

Perhaps it is not that the things themselves have somehow changed to their former condition but that we are somehow in a state of flux and instability and are traveling the multiverses observing the differences without making the connection with our own travels, Since we may be doing this faster than the speed of light, it is incredible that we would be able to perceive any of this, and we would just put down the differences to having been mistaken previously somehow.

A Personal Shift

My blood type is something I am not likely to forget or be mistaken on. For my entire life I have known it to be O Positive, which I have been repeatedly told was the Universal blood type so I would be an ideal donor. Now I am reading that O negative is the Universal blood type, and to my great surprise at least in this dimension/Earth it has always been. I clearly recall conversations with nurses and lab technicians about this subject and their comments on the universal blood type being O positive. Sometimes the memories are so strong and yet now I find it is all too much.

I wish I could just go back to the other dimension/ Earth situated on the exterior Sagittarius arm of the Milky Way. There are many of us that feel even with some things obviously merging that we just don't belong here on the Orion Spur reality/dimension/Earth. Quite a few people have noticed complete changes in family and friends in which they are just not the same people we used to know. Yet we are at a loss for what to say or do now.

Some of them have shifted to this reality/dimension/Earth with us and they also see the strong irrefutable differences.
□

If these differences where only noticed regionally or nationally, other theories on what is actually happening would make sense; however this experience is worldwide. The perceived changes are being reported and discussed in just about every country and by a broad spectrum of individuals from all walks of life.

So how do you handle the day to day new "discoveries?" What do you say, if anything, to waking up to even more things that were established and reliable facts of your existence that seem to have just morphed into something completely different and no one but you realizes this? What do you do when what was so real is not even a factor in other's reality anymore?

Some people wonder if our world was only not destroyed as prophesied in 2012, because those in power used technology to create a shift in our reality/dimension/Earth as prevention of an (ELE) -- Extinction Level Event that was averted.

The Mandela Effect is definitely multifaceted and it is essential that all of the possibilities be considered. Truly we all have to find our own truth. All the wild speculations will terminate. I believe at the end only real truth will survive.

Perhaps we are the new minority who are able to perceive the variations. Maybe it is up to us to envision and then form an original reality. ☐ This may be the time of our being able to harness our misplaced energies and direct them toward seeing and manifesting a better world for everyone.

I'm not certain on any of these unique theories, however we must consider all possibilities at all times - as long as we have considered that they are significant enough to be a real probability somewhere.

Could it also be some type of government operation such as the MK Ultra Mind Control programming working in combination with a subliminal messaging trigger? By use of a "prompt" or "keyword"-- a well-known device in this type of brainwashing, the mind would only recognize the program that has been implanted inside the individual, thus activating only what was programmed to be observed and experienced.

Then there is the idea of other worlds. I have believed in the existence of parallel universes and multiple versions of the same person existing since I was a small child. I had no frame of reference, so I just termed it "here" or "there," and called the different people "the other you."

 I watched things alter and change in my own family and life and learned very quickly as a child NOT to speak up or tell anyone what was happening. I was spanked many times before I realized that what I was seeing and reporting was not seen by others and that my verbally confessing these experiences caused me to get into trouble.

Sadly it still seems as though people aren't willing to share contradictory opinions or views without hostility or name calling. It appears that people can no longer converse in a deferential and informative way. Yet those of us who have had some experiences as children understand often it is best to remain silent to avoid ridicule and we often continue to practice silence as adults.

Just like when we were children we PRETEND that everything is alright, that everything is normal and that nothing is different and that the people who are like strangers to you are still the same. You live on hope every day. You continue to hope that eventually it will all be ok and "back to normal" and that the people you love will be like they used to be. This is the one way that many people are able to function KNOWING that something is "wrong" and "different" and continuing on as though nothing has happened.

In our daily lives we cannot truly say that we have never done something one day that we have done hundreds of times before and still asked ourselves whether we are doing it correctly because it seems somehow unfamiliar. Why would we need to question ourselves?

Having experienced other alternate memories and realizing that I wasn't alone – and that many other people have shared the same alternate memories was just the starting point in my research.

It would appear that we will never be able to understand it all - unless we each experience it subjectively for ourselves. This is what may indeed be occurring.

I find that no matter how much I wish to just ignore this and continue on with "life", I am constantly reminded. Almost every person I have met eventually ends up telling me in some manner about their "close call," or "the time I almost died," followed by how everything in their life "changed after that." Most of the people I have met seem to be unaware of the possibility that they are actually "dead;" at least in their old reality.

For me I know that this is not the same time stream dimension/reality/ life that I experienced previously. I can remember exactly when I probably died recently as well. However, since I've had several of these events, (as I discussed in my first book on dimensions and quantum physics) I have learned more over time.

From countries in the wrong places, and previously deceased people unaware that they are "supposed to be dead," -- to being unable to write a paragraph some days without being constantly autocorrected; everything is different and strange. Therefore in my thinking, either I switched dimensions or universes or realities, or my original location was greatly altered. It could have profound implications for understanding ourselves and understanding what being "real" means. Something unexplainable has transpired. For many people the world seems to have transformed.

The only thing I knew for certain was that I didn't have that bad of a memory of my own existence. My recollections were precise and heavily detailed, and I was in severe shock on seeing things that I had already lived perceived differently. Yes, sometimes they were small things and it seemed unimportant, even minor but those little things still mattered to me and to those who remember differently. Once something is changed in the past, it changes the future. Entire things will be completely different; not partially different. I have noticed that almost all the things that are changed are very public things that would go unobserved with the exception of ordinary regular people who have read, saw or utilized these things on a daily basis, and so would definitely have detailed recollections that would be hard to just ignore.

These are some things I have noticed that are completely different now and for which there seems to be no one fixed explanation.

Muhammad Ali death (early 2016) but I recall his 2009 death, and the funeral on tv.

Frank Gifford died in 2012. His 2015 death was actually quite a surprise.

I Remember Betty White's death in 2014, It was announced that she died—"the last of the Golden Girls is Gone." Imagine my surprise that she is alive and well in 2016.

I remember Nelson Mandela having died in South Africa in prison and His wife Winnie Mandela later became the first black female president of South Africa.

I remember Scotland as being separate from England - its own island and was in the North Sea.

I remember having conversations with a family member that other family members say doesn't exist. Funny thing is he was at my wedding and brought a great wedding present.

Some people myself included remember Ronald Reagan dying in 1999, when he died again in 2004 in this 2015 time stream.

I learned in school that 2 bombs were dropped on Japan: Hiroshima, and Nagasaki however in other timelines it was 3 bombs.

I remember the death of Whitey Bulger in 2013, and a documentary on his life, yet in this time stream, he's still alive.

Mongolia is a gigantic country on global maps and a major world player in many people's reality. For me Mongolia was a province in China. Now in this reality time stream/dimension it is a large country between Russia and China.

I recall the company name as always "Proctor and Gamble" not "Procter and Gamble."

The comet that was the talk of our lifetime was called Hailey's Comet- Here it is Halley's Comet.

The Forrest Gump movie I saw when was first released is totally different than what we have here.

The hit song "Straight Up" by Paula Abdul sounds completely different today than I remember it when it was released.

There was actress Doris Day's dying in the late 2000's yet she is still alive now.

Gone with the Wind's Scarlett O'Hara's famous line: "Wherever Shall I go; whatever shall I do," is here stated differently.

Both my daughter and I remember Cheesecake Factory restaurant meals jokes about the food being horrible and the joke was you just ordered the food to stave off the sugar shock from the magnificent desserts. However, in this time stream it is considered a great place for a good meal and people rave about the food.

I remember being taught about the 52 states. Some children born after 1990 seem to all remember 50 states.

I recall John Goodman's death from a heart attack shortly after the Flintstones movie in 1994, but in this time stream he has lost weight and is alive in 2015.

I remember Korea being SOUTH of China near Vietnam, certainly not out North next to Eastern Russia.

I remember that the Lindbergh baby had never been found, yet here it is reported as having been found dead.

The movie line in the Wizard Of Oz was "Toto, I don't think we're in Kansas anymore." In this time stream it is "Toto, I have a feeling we're not in Kansas anymore."

Reba McIntyre is spelled McENTIRE now- which is different from the Scottish ancestry of McIntyre in my remembered time stream.

I remember Vladivostok being much more North in Russia and not bordering Korea as it is shown on maps in this time stream.

I remember watching on television that event in Tiananmen Square in which the Chinese young man refused to get out of the way of the army tank and was run over and killed. It was shocking and everyone was speaking about it. In this time stream that never happened.

Some remember Fruit Loops breakfast cereal in the 1960s, yet I only remember it first appearing in stores in the late 1970s. And it was always spelled Fruit Loops, not FROOT LOOPS as in this time stream.

I remember the air and fabric freshener product Febreeze, in this dimension it is Febreze.

In my original time stream/ dimension there was no animal called a Narwhal. This is apparently a whale with a long horn on its head like the unicorn. In this time stream it is called "the unicorn of the sea." Growing up I watched nature programs on TV–Jacque Cousteau , Marlin Perkins, David Attenborough, and others and never heard of this. I thought it was some kind of a joke since I've never heard of nor seen a picture of a narwhal before 2016.

I recall Easter Island as having been discovered by James Cook/ Easter Island, and I remember him finding it uninhabited. Rapa Nui is the name of what I knew as Easter Island, given to it by its native people, who have continually inhabited the island for nearly 3,000 years in this time stream/dimension.

I remember the sun being bright yellow, not white, and I learned in science classes at school that there were only 4 or 5 cloud formation types , but in this time stream clouds appear in odd shapes and forms and there are over 20 types here.

Thanksgiving was always on the third Thursday of November in the United States. And in this time stream, it's the fourth Thursday in November. It stands out in my mind because my grandmother taught it to me as a child, when I learned the countdown to Christmas day.

I remember the peace sign become popular in the 1970s; it had the arms facing upward; never downwards as in this time stream.

I remember the GREAT Pyramid of Giza being off into the desert MILES away not literally 700 FEET from the suburbs of the city of Cairo Egypt, as it is here.

I remember that Jane Goodall died and was remembered for her research on gorillas, when in this time stream she is still alive and famous for her research with chimpanzees.

Gorillas in the Mist was a movie which had a TV premier and I distinctly remember the movie I saw was about Jane Goodall; staring Susan Sarandon. In this time stream Sigourney Weaver is the actress in that movie and it is about Diann Fossey.

I remember the pictures of this massive white statue called Christ, the Redeemer overlooking the city of Rio de Janeiro on a gigantic white rectangular base. Now it is just a large statue. The base has also radically and mysteriously changed to a smaller base and is a black square cube.

I remember a BBC America Television show called MI-5, however in this time stream it is called SPOOKS, and while still about MI-5, it was never called that especially in the US.

Cartoons were Looney Toons now Looney Tunes and Merrie Melodies in this time stream, yet I knew it as Merry Melodies my entire childhood.

I remember a peanut butter known as "Jiffy" the original brand name. So when I saw "Jiff" peanut butter I thought it was a name change by the company. It seems that at least in this time stream there was never a name change and it has always been known as "Jif." However I remember my brother and I being very insistent with my mom when we were children that she only buy "Jiffy" and not "Skippy" another brand of peanut butter. I even remember the song from the commercial.

I remember Oscar Meyer as a deli product company, in this time stream it is Oscar MAYER. I even remember singing the song in the commercial…about "my bologna has a first name it's ----O-S-C-A-R, my bologna has a second name it's----- M-E-Y-E-R….!"

I also recall that all traffic lights were green yellow and then red at the bottom, so I was surprised when I noticed it in reverse.

I also remember the spelling of words being completely different. I spelled a word as "suprise" now it's "surprise" and "lightening" instead of "lightning", and "realise" is now "realize." I was always big on reading and writing and had entered spelling contests every year as a child. I paid attention to words and I am a writer now, so I find this bizarre. We were taught the proper grammatical usage is "my brother and I," now here in this time stream it is "my brother and me."

Here combined words are non-existent: 'infact' is now "in fact"; afterall to "after all"; overall to "over all" moreso to "more so;" alot to "a lot"; alright to all right; and no-one is "no one."

"Dilemma" is remembered as being spelled "dilemna" and "dammit", as "damnit" The spelling of the nation of Columbia changed to Colombia.

The colors chartreuse and puce have switched here. I remember Chartreuse a pink -reddish purple, not puce's yellowish-green color.

The automobile symbols are different also. Volkswagen – VW, here has a space between the two monogram letters, and Volvo in this time stream has an arrow added to the circle, making it the symbol for "male,"and not the circle missing a piece that I remember.

Vancouver Island seems larger here and British Columbia is much larger also on these maps.

I remember New Zealand being one land mass. In this time stream it is now two islands and it is bigger than Italy.

The Bahamas were NEVER just off the coast of Florida in my time stream, only Bermuda was. Cuba was NEVER that close to Mexico. Also there was no island off the coast of Cuba!

When I visited NYC years ago Manhattan Island jutted out into the Atlantic. The statue of Liberty was on an island a little farther out into the Atlantic and not near New Jersey. You had to take a ferry to get to Staten Island as they never had a bridge.

Here in this time stream I learned there are 4 bridges to Staten Island. I had no idea that there was any bridge. I always thought that you had to use a ferry to go to Staten Island.

I do remember in the movie Working Girl, actress Melanie Griffith had to ride the ferry back to Staten Island and I clearly recall the scene. In this time stream the movie does not have that scene.

Martha's Vineyard was a district on Long Island. It has been moved away, leaving the Bay in Long Island, here Martha's Vineyard is an island.

I recall Sri Lanka being directly South of India, not off to the East of it. I was shocked to see Gibraltar moved from the strait between Spain and Morocco –to be on the East coast of Spain.

I was stunned in particularly by South America's 1000 mile eastern shift, out of what I recall as the straight alignment with North America.

The JC Penny Store in this time stream is JC Penney.

American Television chef of "Bizarre Foods" was Andrew Zimmerman, here he is Andrew Zimmern.

The host of the Twilight Zone television series was known as Rod Sterling, here he is Rod Serling.

Walmart was ALWAYS a blue logo- never Wal*mart in white logo in my original dimension.

The Talladega Superspeedway and the Daytona Raceway were in Florida, however in this dimension the Talladega Superspeedway is in ALABAMA. I was shocked that there isn't even a town called Talladega in Florida.

Then again there is that Rock of Gibraltar. It is British owned. It is in my time stream/Dimension, a source of contention between England and Spain because Spain believes due to its proximity to their coast it should be considered as Spanish territory even though it is an island offshore. This is how I remember it; an island of disputed territory, not a part of the land mass of the country of Spain, sitting surrounded by water facing Morocco.

I vividly remember a land mass being called the North Pole it was never a large lake. New Zealand was above Australia and it was one land mass and not in two pieces.

Australia is now half the size I remember and is missing part of its shape at the top. Indonesia and Australia are much closer to each other.

Since Quantum Mechanics states that every eventuality is played out somewhere in the universe, we could have entered into an area with different eventualities. There is consciousness in things and areas in which perhaps we have never considered. Could this be impacting our reality. The basic structures of our own brains are simply bands of energy. This energy is the basis of everything. Modern Quantum physics asserts that this energy is not dormant or silent it is pulsating and it is aware. There was a perceivable moment in which everything changed. Things looked much different before that time. The idea that there has been some strange paradigm shift and the world you think you were living in and the information that you embrace as fact was not the only fact but another version of those facts is quite sobering.

My memories are for me the only real thing I have left to judge my shifting dimensional/ realities by, so I defer to this whenever I detect a change.

Then there are those people who deny that this phenomenon is a worldwide one. They are wrong. Growing up in a multicultural setting allowed me to interact with many people from various part of the world. My own family is quite diverse, so I am able to gain a broader spectrum of observation through this as well.

The Indian singer Hans Raj Hans received one of the highest awards from the Indian government; the Padma Shri. He sang Punjabi and Sufi music and was credited for its revival and preservation. He was also a professor of music at a University in India. He made videos and albums and was very much respected and quite conservative.

In this dimension he is involved in politics and was elected to public office in Punjab India. He is also in a video wildly dancing while singing the song – "Tote Tote Ho Gaya Bichoo" (Hindi) with some woman who looks like an Indian version of Cher. (Sonny and Cher).

He sang one of my favorite songs- "Ae Jo Sili Sili Aundi Ye Hava",(Hindi). I had all of his CD's. I was shocked to find that in this dimension there are CD's that are really old, but I have NEVER heard of them.

He was a guest at my father's house when he toured and gifted my Dad some CD's that I gratefully received. So I was very familiar with all of his work.

In this dimension he is singing and dancing in a video "Dil Chori Sada Ho Gaya" (Hindi) only in my old dimension this was a big hit for another singer-- Harbajan Mann. I KNOW he did not sing and dance in this kind of music, or that song and he never dressed like that.

Haifa Wehbe is the Lebanese singer (former Miss Lebanon) famous for a song, now sung by someone who didn't exist in my timeline- Marwa ?? Here she is apparently a big Lebanese singer famous also, except she was nonexistent in my dimension.

There are two Spanish Soap Operas "Gitanas" and "La Mujer en el Espejo" that have completely different story lines and endings in this dimension. I can remember watching one particular episode years ago that was a cliffhanger that NEVER happened here and is not on our DVD's or even our friends video tapes..

So yes, it is a global phenomenon, at least for quite a few of us it is.

I remember anatomically the basic set up of the human body. As a former pre-med student, anatomy was not a subject I could easily forget. I know that we had two floating ribs on each side of the lower rib cage. Here in this reality/dimension we have no floating ribs in the front but have a pair; one on each side in our backs. The intestines were not an untidy massive muddle starting up in our chest area. It was neatly placed under our stomach and our liver was smaller and at our lower right in our back. Additionally our stomach was located lower behind our belly button in the abdomen area, was placed horizontally not vertically, and it certainly wasn't up under the ribcage and to the left. The pulse was always detected in the middle of the wrist, and not on the side as it is here. The heart was bigger and slightly to the left in the upper chest area and not in the middle of the chest as it is here. Right below and behind the stomach were the kidneys and the liver. The spleen was the largest organ, next to our skin. Now the kidneys are the largest organ and are sitting on top of the vertical stomach up under the ribcage. This peculiar placement of organs in the human body now seems to leave very little room for the lungs. I also only remember us having twelve ribs, and not the twenty-four that people have now. This is not the human body as I knew it to be.

The night sky has changed tremendously. In 2011 my stepson Wally and I used to stare into the night sky night after night for weeks and marvel at how many things we could see that we just were unable to identify. We wondered about the strange behavior of the moon, and thought that perhaps the night sky we were watching might be nothing more than light projections on a Reflective Dome. Then there were times I could clearly see two suns in the sky.. I grew up having learned that our earth was on the outer arm of the Milky Way spiral. We are in a solar system that's part of the Milky Way Galaxy, but it seems we are on the Orion arm of the spiral which is way inside, and no longer on the Sagittarius Arm.

To my memory the sun shines quiet differently and it emits a light like a concentrated ray not diffused or soft. The sun is no longer YELLOW but WHITE and now it can burn you immediately. I remember when I was growing up the sun was a golden yellow and not a searing bright and white fire ray. Orion's belt is almost as bright the moon. I remember when it was rare to be able to make out the Pleiades in the night sky and that was with assistance. Now we can see the entire Milky Way. Today the Pleiades are very bright and right overhead, where previously you had to use a really good telescope, and it seems that all the stars are in a different position. I personally have given up on telling my friends and family (those that don't see any changes) about the Mandela Effects and the shifting of reality and dimensions. I just document and write about what I have learned to help others like myself understand what may have happened and what may occur next.

This book is the fifth in the series and as the Mandela Effect seems to be increasing; our ability to cope with all that is affecting us also seems to be evolving. A guide to what this all encompasses seems needed somehow, and I hope this book can somehow serve as one.

The world as we know it can and will change in an instant. There are many people I believe are randomly moved to nearly identical universes, taking the place of their alternate self without ever realizing exactly what has happened. They may feel strange and some may even notice things that are not quite right, but without ever realizing that it could be something else just dismisses the occurrence or feeling as insignificant although puzzling. I had no idea I was not the only one. I thought this was just happening to me until I discovered that there are other people experiencing this as well.

I've always felt that humans are spiritual beings having a physical experience; the Mandela Effect seems to validate it for me. I can't reject my memories and make-believe that this is the reality/world dimension I have been living in, this is not the reality I remember.

One cannot prove nor disprove the Mandela effect. The whole principle behind the phenomenon is our reality has one way or another has changed so it doesn't matter how far back in time you look, the present reality will reflect what was true for its dimension/reality.

You are somewhere else, and in this new place this is the way they do it here. When well-meaning but frankly senseless individuals say dismissively, "Your memories are wrong, it's all a 'psychological operation,' or you are 'being misled by Satan,' or you just 'didn't understand geography', or 'didn't know how to spell,' or 'really had not read your Bible" it is trivializing and contemptuous showing lack of understanding and any comprehension into something many people are experiencing. So the best idea is to be very selective with whom you share any of your noticed revelations or discoveries, if you do not want to face either being ridiculed or dismissed as "strange."

Different people pay attention to different things, so we'll not all be similarly overwhelmed by identical changes to things we knew of in the past, and occasionally it takes some time before others may realize that something strange has been going on. Because someone says something negative, does not mean it's true in your life and cancels your remembrances.

When you first realize that reality is not what you always thought it was it is personally distressing. Remember some people may be having the same experiences, yet denying it due to their religious beliefs, or their educational indoctrination. Not every dismissal is personal; some people may be struggling with this privately and are angry and frightened.

I believe the next phase of this experience will be practically overnight physical changes to our perceived environment and our society that will be so enormous the general population will begin asking questions and demanding real explanations.

The sad thing is that there are humans just like ourselves who know exactly what is happening and do not care if we are traumatized and distressed. They have no capacity for love, or kindness or empathy, and as is to be expected will meet their appointed end. You see the universe has laws too.

Since we are unable to do anything to stop all these mysterious changes, it helps to agree that they are indeed occurring, that they are important. I believe that our Creator, G-d is still in control and we are not alone. I still believe.

In every modern court of law the witnesses MEMORY is held as valid testimony. And usually one or two witnesses are sufficient. In this reality /dimension/ universe change experience we have more witnesses than would be required in a court of law to testify and have our MEMORIES be accepted as the truth. The fact that so many bizarre things are taking place in this reality/dimension means that reality is not what we've expected it to be. It helps to remember everything is real somewhere.

Chapter Three

The Soul And Its Confusion

"To be moved confuses the soul. One cannot convey these kinds of memories any more than the events of a dream...

...if I have complained too long, it is because my memory, no longer having any fixed abode, has to carry its luggage with it." -Jean Cocteau, The Difficulty of Being

For people who don't know that this is a different place, just reading and seeing the bible scripture changes here is a major shock. Those that do see what has happened understand that the things in this reality weren't just changed or altered , but that we are probably from a parallel reality/dimension that has some major differences from what we have known in the Bible.

Well, this is where having the scriptures stored in your memory becomes invaluable. We have moved into this time in this reality/ dimension now without most people really noticing anything different or being aware of the relocation unless they had memorized the Biblical words and to their shock found them changed from what they have known.

What I clearly remember regarding these examples does not exist in this present reality/ dimension. A single word or often the position of words has changed. If you searched the scripture sites on-line or searched in any Bible-- even your personal copies of many years, the scriptures are completely different.

This may have been done when our old reality merged with other time /reality/ dimensions in which the altered biblical text had already been in existence there. This dimension would have to be much further away from our original one, for there to be so many obvious differences that are easily noticed. The idea that there could be multiple versions of us is easier to accept than multiple version of the KJV Bible for many people. The idea that there has been some strange paradigm shift and the world you think you were living in and the information that you embrace as fact was not the only fact but another version of those facts is quite sobering. Additionally it takes a large leap forward in thinking to consider that the fabric of reality shifted at some point in the past, and we are in a parallel inhabitable reality/dimension/universe. It even seems that we are continually moving between them.

The way I see it the only proof we have, if WITHOUT A DOUBT WE ARE IN A NEW REALITY/DIMENSION, are the memories of those of us who for whatever reason shifted to this reality/earth/dimension/universe, and kept those old memories. I have read the entire Bible, including both Apocrypha many times in thirty years. I have also studied various translation of the Bible due to my work and research. I have read all of the Old Testament and pseudepigrapha, apocrypha and sacred writings along with most of the surviving New Testament apocrypha and the gnostic gospels including the Lost Books of the Bible.

Nothing could have prepared me for this drastic change in what I knew as the bible. It took me a while before I let go of the fear that I would daily find more changes to what for me has been a beloved and treasured book.

There are as far as my experience at least two versions of this book and they both are supposed to be considered correct. Is it possible for each dimension to have its own version of the Biblical text? Who is right and who is wrong? Would it be wrong to cling to the version I have memorized from **before** the shift? Or am I expected to adapt to the version here *after* the shift, that the believers here claim has always been this way here?

Thankfully my faith does not center on a book. Yet the reality of what is happening is so hard to comprehend. The proofs are available for people with "eyes to see and ears to hear."

There are now multiple verses in the Old Testament using the word "matrix," in place of the original word "womb." In this dimension/reality they did not however replace all instances of the word womb in scripture.

Exodus 13:12 That thou shalt set apart unto the Lord all that openeth the **matrix,** *and every firstling that cometh of a beast which thou hast; the males shall be the Lord's.*

Exodus 13:15 And it came to pass, when Pharaoh would hardly let us go, that the Lord slew all the firstborn in the land of Egypt, both the firstborn of man, and the firstborn of beast: therefore I sacrifice to the Lord all that openeth the **matrix***, being males; but all the firstborn of my children I redeem.*

Exodus 34:19 All that openeth the **matrix** *is mine; and every firstling among thy cattle, whether ox or sheep, that is male.*

*Numbers 3:12 And I, behold, I have taken the Levites from among the children of Israel instead of all the firstborn that openeth the **matrix** among the children of Israel: therefore the Levites shall be mine;*

*Numbers 18:15 Every thing that openeth **the matrix** in all flesh, which they bring unto the Lord, whether it be of men or beasts, shall be thine: nevertheless the firstborn of man shalt thou surely redeem, and the firstling of unclean beasts shalt thou redeem.*

More Differences

*Genesis 3:15 **"crush" and "bruise"** "And I will put enmity between thee and the woman, and between thy seed and her seed; (he instead of it?) it shall **CRUSH** thy head, and thou shalt **BRUISE** his heel."*

*This is now changed to **"bruise" and "bruise"** "And I will put enmity between thee and the woman, and between thy seed and her seed; it shall **BRUISE** thy head, and thou shalt **BRUISE** his heel."*

Luke 20:24 ***"Denarius" to "Penny"*** "Shew me a denarius. Whose image and superscription hath it? They answered and said, Caesar's."

This is now changed from *"**denarius**"* to *"**penny**"* "Shew me a penny. Whose image and superscription hath it? They answered and said, Caesar's."

This verse in Revelation is also changed to *"**penny**"* from *"**denarius**."*

Revelation 6:6 "And I heard a voice in the midst of the four beasts say, A measure of wheat for ***a penny***, and three measures of barley for ***a penny***; and [see] thou hurt not the oil and the wine."

Mark 13:10 *"**preached**"* to *"**published**"* "And the gospel must first be *"**PREACHED**"* among all nations."

This verse is now also changed. "And the gospel must first be *"**PUBLISHED**"* among all nations."

Matthew 26:45 Jesus commands them to wake up." Then cometh he to his disciples, and saith unto them, "***awake, from your rest***" behold, the hour is at hand, and the Son of man is betrayed into the hands of sinners."

This verse has been changed to Jesus commanding them to sleep.

"Then cometh he to his disciples, and saith unto them, ***"Sleep on now, and take your rest"***: behold, the hour is at hand, and the Son of man is betrayed into the hands of sinners."

Luke 22:20 Changed from ***"New Covenant"*** to ***"New Testament"***

Luke 22:20 Likewise also the cup after supper, saying, This cup is the ***new covenant*** in my blood, which is shed for you

Now it states in Luke 22:20 Likewise also the cup after supper, saying, This cup is the ***new testament*** in my blood, which is shed for you.

Acts12:4 Changed from ***"the Passover"*** to ***"Easter"*** in the KJV.

Now it states Acts 12:4 "And when he had apprehended him, he put him in prison, and delivered him to four quaternions of soldiers to keep him; intending after **Easter** to bring him forth to the people."

BEFORE Acts 12:4 "And when he had apprehended him, he put him in prison, and delivered him to four quaternions of soldiers to keep him; intending after ***the Passover*** to bring him forth to the people."

BEFORE John 3:16 KJV "For God so loved the world, that he gave his only begotten Son, that whosoever believeth in him **SHALL** not perish, but have everlasting life."

AFTER John 3:16 KJV "For God so loved the world, that he gave his only begotten Son, that whosoever believeth in him **SHOULD** not perish, but have everlasting life."

BEFORE Luke 21:9 "But when ye shall hear ***of wars and rumours of wars***, be not terrified: for these things must first come to pass; but the end is not ***forewith***"

AFTER Luke 21:9 "But when ye shall hear **of wars and commotions**, be not terrified: for these things must first come to pass; but the end is not **by and by.**

BEFORE Genesis 9:16 "And the **RAINBOW** shall be in the cloud; and I will look upon it, that I may remember the everlasting covenant between God and every living creature
of all flesh that [is] upon the earth."

AFTER Genesis 9:16 "And the **BOW** shall be in the cloud; and I will look upon it, that I may remember the everlasting covenant between God and every living creature of all flesh that [is] upon the earth."

BEFORE Galatians 4:25 "For this **HAGAR** is mount Sinai in Arabia, and answereth to Jerusalem which now is, and is in bondage with her children."

AFTER Galatians 4:25 "For this **AGAR** is mount Sinai in Arabia, and answereth to Jerusalem which now is, and is in bondage with her children.

My memory is that the Book of Revelation ended with Revelation 22:20, which is: "He which testifieth these things saith, Surely I come quickly. Amen. Even so, come, Lord Jesus."
Now there is a **NEW VERSE ADDED** Revelation 22:21 "The grace of our Lord Jesus Christ be with you all. Amen." (Protestant Benediction?)

BEFORE Ephesians 3:19 "Now unto Him who is able to do exceedingly abundantly above all that we ask or think, according to His power that worketh in us. 20 **To Him be the glory, honor, and praise forever and ever. Amen**"

AFTER Ephesians 3:19 19 "Now unto him that is able to do exceeding abundantly above all that we ask or think, according to the power that worketh in us, **20 Unto him be glory in the church by Christ Jesus throughout all ages, world without end.** Amen"

The Catholic word **BISHOP** has now replaced **ELDER** in the KJV.

1 Timothy 3:2 "A **bishop** then must be blameless, the husband of one wife, vigilant, sober, of good behaviour, given to hospitality, apt to teach;"

Acts 1:20 "For it is written in the book of Psalms, Let his habitation be desolate, and let no man dwell therein: and his **bishoprick** let another take."
Bishoprick replaced office.

Philippians 1:1 "Paul and Timotheus, the servants of Jesus Christ, to all the saints in Christ Jesus which are at Philippi, with the **bishops** and deacons:"

1 Timothy 3:1 "This is a true saying, if a man desire the office of a **bishop**, he desireth a good work."

Titus 1:7 For a **bishop** must be blameless, as the steward of God; not selfwilled, not soon angry, not given to wine, no striker, not given to filthy lucre;

1 Peter 2:25 For ye were as sheep going astray; but are now returned unto the Shepherd and **Bishop** of your souls.

Additionally the names of the Four Gospels are quite different in this dimension/reality/Earth. They are here written as Saint Matthew, Saint Mark, Saint Luke and Saint John.

1 John 5:7 and 8- 7-For there are three that bear record in heaven, the Father, the Word, and the Holy Ghost: and these three are one. 8And there are three that bear witness in earth, the Spirit, and the water, and the blood: and these three agree in one.

It now states- 7For there are three witnesses; 8 the Spirit, the water, and the blood—and these three are one.

Daniel 8:24 His power shall be mighty, but not by his own power; He shall destroy *fearfully*, And shall **prosper and thrive**; He shall destroy the mighty, and also the holy people.

It now states here-Daniel 8:24: His power shall be mighty, but not by his own power; He shall destroy **wonderfully**, And shall **prosper and practice**; He shall destroy the mighty, and the holy people.

There are at least 30 more scriptures in the book of Daniel that are completely different to what I knew before.

I know my own Bible and I know that none of those changes were EVER in my Bible. If they are the original to your bible, and they have always been there, you are from this dimension/reality and Earth. However for those of us who see a difference learning how to handle this can be very difficult. It is very hard to offer any observation on the religious aspects of these phenomena. My belief system does not require an adherence to any written text in any physical book, so for me there is no difference as God is still the same in my mind and understanding. Yet here the vast majority still maintains that the bible is the way it has always been. Nonetheless, a very small minority say it has changed from what we have always known. I find that I am tentative to speak about anything regarding the dimensional shift or any of the examples because often people will instantly talk down the implications or just dismiss the idea completely. I am impatient when it comes to people ignoring reality. Stay strong through the attacks and deniers.

When it comes to the deep beliefs that one holds, that decision will have to be made personally; as it is about satisfying our own souls. At the end of all this, it is still going to be amazing. We are able to witness it all happening before our eyes and also to live it every day. I believe that there are no accidents. God wants us to see what is happening and what will occur next. If we truly have been moved the only thing we must retain are the memories that we know are our own.

Chapter Four
Technology Gone Wild

The scientist at CERN in 2015 requested assistance from computers across the world. Why with D-Wave super computers sitting at the core of CERN, did they want more even more computer power from ordinary people in other countries on the earth? I suspect there was more to this than we were being told publicly. These quantum computers are designed to reach into parallel dimensions/realities and bring back whatever it discovers. There are at least three located at CERN, and Google leases one of them. It is AI with consciousness; so it is self-aware.

It is also able to open parallel universes and dimensions and alter our reality, by its ability to interface with the collective consciousness. Just one of these computers takes only a few seconds to figure out a problem that the most powerful computer on the earth would have taken ten thousand years to calculate. A giant leap of this magnitude in AI doesn't just happen without some outside assistance. Perhaps it was in its construction materials as my research suggests.

The co-creator of the D Wave quantum Computer, Dr. Geordie Rose from Burnaby, BC Canada, has admitted to the computer being able to access parallel universes just like ours. As they do this they are "seeing slight differences between all of these dimensions and universes." He admitted that science has reached the point where "they are able to exploit those other worlds and dimensions."

This quote from David Deutsch, "Quantum Computation... will be the first technology that allows useful tasks to be performed in collaboration between parallel universes," should make it more than obvious that parallel dimensions/realities do exist. You may think that they are referencing technology that is theoretical, but not yet relevant in usage and you would be wrong. There is something within the structure that empowers the quantum computer to gain access to parallel universes and dimensions now.

They have several of these super computers in use at this time. There is one at Los Alamos National Laboratory and one is owned by Lockheed Martin and used by University of Southern California. Another was installed at NASA's Advanced Supercomputer Facility in Silicon Valley.

They are black in color, box shaped and metal, ten feet wide and 12 feet tall in size. They have a refrigerator inside called a pulse two dilution refrigerator. It is a cryogenic mechanism that provides constant cooling to temperatures as low as 2 mK, with no moving parts in the low-temperature area. The refrigeration procedure uses a combination of two isotopes of helium: helium-3 and helium-4. These refrigerators have a device called a pulse tube which emits a type of noise, about once every second that sounds strangely like a heartbeat. Inside these computers is a very tiny chip. This chip serves as the interconnection for more than one dimension to overlap, due to its ability to have more than one value to its quantum bit or qubit. This is something that cannot be done by a regular computer.

As those devices would either be 1s or 0s, but the quantum computer can be both at the same time in essence like being in two "places" at once.

While stored as a string of 1s and 0s, their equals in a quantum system, these qubits remarkably can be both 1s and 0s at the same time. They have also stated that every time the qubits are doubled, they are increasing the number of parallel universes and dimensions that they have access to. They as of this writing they had 2 to the 500th power "living" in the chip within the quantum device.

However they never address any side effects, such as people being thrown around like dice in and out of different dimensions and realities and ending up in parallel dimensions and universes knowing that something is very wrong but unable to understand just what. Is it any wonder that reality is unnatural for thousands of people?

They claim that the shadows of these alternate dimensions/universes are intersecting with our own, and they are able to "grab resources" from those dimensions and universes into this one. What? I wonder what else they are grabbing from other dimensions. Have they become the "Robber Barons" of the galaxy? After all of this are they really not exemplary scientist and pioneers, but rather just criminals and thieves with a better more advanced weapon?

The creators of the D Wave Quantum Computer have been doubling the qubits on the chips giving them access to more and more dimensions/ realities and universes each time, and have at least admitted to doing this for the past nine years.

They have brashly declared that they will continue to do so doubling the amounts of dimensions/ realities and universes that they can access and take whatever they want from.

They claim that their model Vesuvius (2012) quantum computer is 512 qubits and is one half million times faster than the previous model Rainier (2010) at 128 qubits. So they are also predicting that there will be machines that will be self-aware and able to do anything that humans can do, and that these quantum super computers will be what will cause this to happen. What if we ourselves are also being changed in a way to make us compatible with an AI system that will try to replace "consciousness" as we know it?

I must be honest as I write these lines I kept thinking about Skynet. You know what I mean. Yes, the computer in the Terminator movies that gained self-awareness after it had spread into millions of computer servers all across the world. Skynet was a computer system created for the U.S. military by the defense company Cyberdyne Systems in the Terminator movies. After activation it gained artificial consciousness, and the terrified operators, recognizing the full degree of its abilities, attempted to turn it off. That computer comprehending the scale of its own vulnerability when its creators tried to deactivate it turned on the humans. In the interest of self-preservation, Skynet decided that all of the human race would endeavor to terminate it and declared war on the human race. It used servers, mobile devices, drones, military satellites, war- equipment, androids and cyborgs – as a Terminator, and additional mainframe systems in a world war.

**As a side note let's just hope if things go terribly wrong that the quantum computers do not watch the Terminator movies.

Is this going to lead to a convergence of all our multiple selves and dimensions/ realities all merging into one dimension/reality and one super self? And what of the activities at CERN with the LHC and those quantum computers: will they come along as well?

Not all CERN scientists comprehend accurately what they are undertaking at the LHC. Some of these scientists believe they are looking for something called the "Cosmic Tree," which is an elaborate map of light matter and dark matter originating in parallel dimensions.

They most likely do not even know that their superiors have had access to other dimensions/ universes for over nine plus years, and are now setting themselves up to access even more dimensions by their own admission. How could they do something so fantastic as what they are claiming? They received help in a most bizarre form.

There is reportedly a substance termed "Black Goo." This is what is rumored to be the substance used to build the quantum computer giving it "consciousness" and making actually AI. The substance is said to be an intelligent nano-tech conscious liquid. This could be considered a kind of interstellar AI or a "seed device" for generating life inside a new biosphere. It is said to contain remnants of its original lifeforms; spider type beings. It can take the form of an abiotic mineral oil comprising high amounts of m-state gold and iridium.

No one seems able to give a definitive answer to its origins and to the question of its time here in the earth. Some say it appeared here over 80,000 years ago during the Lemuria time frame, and may have contributed if not have been the direct cause of their societies collapse and destruction. However it is all at best speculation. Today we have this substance reappearing all over the world.

In February 2016 a town in Michigan reported that a black tar-like substance showered down on their cars, terraces and driveways and the material still remains a mystery.

The city's fire chief said that it was not bird dung and was not flammable. They still have not found a definite answer and the residents are still awaiting an explanation.

Black Goo is said to be found in the largest amount or deposit in the ground in Paraguay. It is a liquid crystal that looks a lot like black tar, with apparent intelligence and seems to be living, because of its movement under the microscope. It has so far been found in diverse places such as Germany and in Galveston, Texas when their rain water was tested. It was a substance used in pagan rituals in antiquity, because of the force that could be felt in it presence.

There are solid stones made of this and one black stone is said to be that black stone in the Kaaba in Mecca. There is one also used in the black altar in St Peters Basilica in Rome. Both of these stones are said to be remnants of meteorites and have a history of being considered distinctive and extraordinary.

The liquid black goo was also found in the Falklands, on Thule Island and under the Gulf of Mexico. This information was kept from the public and was known only to the people directly working with it, and harvesting it. This same Black Goo was mysteriously found floating on the Indian River Lagoon in Florida's Hutchinson Island Bay in 2013. It soon disappeared after being reported by local residents.

It was also reportedly found at the Deepwater Horizon Incident which is also termed the Gulf of Mexico Oil Spill in 2010. This was the largest marine oil spill in history seemingly produced by an explosion on the Deepwater Horizon oil rig in the Gulf of Mexico on April 20, 2010. No civilians were allowed to work at the clean-up site.

They permitted just military personnel with specialized equipment and those civilians who did come into physical contact with the oil then became sick and some even had to eventually have limbs amputated.

There were potentially 200,000 people who may eventually have to take advantage of a proposed medical settlement with BP regarding forthcoming adverse health effects linked to helping in the cleanup and unknowingly being exposed to more than petroleum oil. As usual the real-world evidence of all the people with chronic medical issues and the massive environmental ruin that resulted from that disaster is still growing.

There are apparently two types of the black goo. One is coming up from the inner earth and the other has somehow come from outside our earth.

The one from the inner earth is termed "Sentient Oil," and is said to produce empathy and a feeling of love in those in its vicinity. When in the vicinity of the "Meteorite Type Oil" all feelings of empathy disappear and a feeling of cold fear is felt. It is said that it has the ability to enter DNA and be able to modify it.

This substance; and we are not sure which one, but we can guess, has been used in the creation of these super computers, because it is said to be programmable and to be able to self-generate. It can also be used to program a living thing. So that the use of this in the construction of a computer could conceivably be that external assistance enabling our technology to make that massive leap almost overnight to a new AI quantum machine. And yes the leap into AI quantum computers was done within the construction materials itself and not necessarily the software or programmers.

The activities of the people using the quantum computers and the undertakings of the LHC at CERN could be quite possibly the method in which many of us have found ourselves in an alternate dimension/reality/ universe with apparently no way to return to our original one if it still exists. I believe there have been many shifts/ changes with our reality. As vibration energy at certain frequencies can affect particles causing them to separate or to change shapes, we know that movement is one of its effects. It is only a matter of "time" before various particles can exchange locations spontaneously or with assistance.

Actually for many of us that may have already taken place and we are in a different reality/dimension/earth than our original one. I believe that the LHC at CERN in tandem with whatever D Wave quantum computer they have searching the universe starts blending with every other dimension that has a CERN and an LHC doing the exact same thing at the exact same time. They would have to actually synchronize for greatest effect.

These are similar realities/dimensions yet they are not the same. The dominant frequency and the place on the earth where the shadow of that dimension overlaps with your current reality when the power is released, probably determines what set of people in our original reality are relocated dimensionally and exactly which reality and dimension they are sent into. They become "CERN LHC refugees" without even being aware of what has actually transpired.

I personally have given up on telling my friends and family (those that don't see any changes) about the Mandela Effects and the shifting of reality and dimensions and now I just document and write about what I have learned to help others like myself understand what may have happened and what may occur next. Now after years of researching this subject, it's exceptional for anything to take me completely by surprise, but it gets more bizarre every day. I believe that the Creator G-d can stop the activities at CERN with the LHC, and the activities with the quantum computers and dimension shifting anytime He chooses to do so.

I also believe in Universal law which requires certain things to just continue to play out to their expected end. Perhaps these actions by these evil minded men are the equivalent of roulette or craps they are attempting to play with the Creator and the Universe itself. Foolish people don't they know the house always wins.

I can't deny my memories and make-believe that this is the reality/world dimension I have been living in, this is not the reality I remember. I have no doubt now that parallel dimensions exist, whether or not they explain the Mandela Effect is another matter. Ignorance is happiness for many people. This is because most people do not want to know the truth, if they acknowledge the truth then perhaps they will need to change their views. The physical changes to the earth, the geography and the astronomy coupled with the changes to the reality that you have always know are somehow tied together.

I suspect from all the evidence that we do live in a multi-universe, and that the LHC at CERN along with the other technologies working in tandem, may have already destroyed many dimensions. This resulted in the fact that we have somehow moved completely from our original location. The idea that the use of at least three D-Wave Computers simultaneously in this dimension, probably timed for greatest effect with the same event in other dimensions is to my mind the most likely catalyst in the changes.

From my perspective, I am just appreciating the experience of living now, and trying not to get overcome by the new changes that are in my face on a daily basis.

Our ignorance is still our weakness and to gain strength we must learn and share what we know. We do know that in the face of all of this the world media empire has been silent. With budgets larger than many nations the entire empire is owned and controlled by a mere handful of people that we never see. Their power and control is unbelievable. Media's many forms are simply propaganda tools to instruct people in what to do, whom to believe and of course whom to fear and obey. They are so powerful that they create and bring to life political agendas and economic policies. They are even able to influence us to send our precious children to fight in wars, that they themselves have already selected by declaring which nations are to be vilified and invaded and when. Do you believe that with this much power and control, they have no idea that there had been dimensional changes and vast segments of the population have noticed it? Every news item that is broadcast is designed to program us toward their agenda. Now they are silent. Why?

Regardless of where this energy initiates or why it is now here, the bottom line is that it is happening and we are all affected by it and it is increasing every day. The greater our worldwide consciousness becomes, the easier it is for us to ask for and receive our demands as a collective group.

Once the critical mass of people on the Earth shares this, there will be no confrontation. I personally believe that we are gaining that critical mass and we are at the point of change for this earth and its people.

Chapter Five

What Do We Do Now

If you particularly remember events the way they transpired here, you have totally no reason to believe that you are from a different "earth/dimension/reality. Most people only realize something is wrong when the find themselves faced with at least one discrepancy, that they could either not ignore or explain to their own satisfaction.

A portal is a two-way inter-dimensional entrance or doorway into numerous realities including parallel universes/ earths/ dimensions. Between these dimensions/universes/earths are curtains or veils that are undetectable to us in the physical way, but isolate us from other dimensions/universes/earths. These are border guards or boundaries a form of energy and frequencies between planes of existence that keep us distinct from each other. Within each of these curtains or veils are doorways or entrances that can be opened to allow access between any of the dimensions.

Since those of us who have many differing recollections about the same topic, understand that this may be from having moved through several dimension/earths/realities where we remember events there, and we believe that this is happening throughout the Milky Way galaxy and throughout the Universe; it's probably also happening throughout multiple dimensions and with various versions of ourselves as well.

I have a whole lot of completely distinct memories of versions of the past which are no longer true; at least in this reality/dimension/earth. Individuals tend to side with physical evidence, even when their memories are certain and they know that their reality has changed. You cannot just let go of your memories and perceptions since those are the methods by which we in fact interact with reality.

But what if the people in your own family; your wife and your children are like copies of your old wife and children. You know that they are not the same and you wonder where are the "real" people that you loved and lived with all along. But the thought is so crazy that you have to dismiss it to maintain your sanity. If this seems outlandish to you, it is because your personal experiences with reality collapsing into another similar realm have not led to enough memory discrepancies to lead you to the same conclusion.

This is an amalgamated reality. Unbalanced realities have merged, due to the massive technological interferences and memories are all we have left that serve to preserve information. The FACT of the matter is that things that are strange tend to stand out, particularly when you see them over and over. I think that we are conditioned to chalk up these reality shifts as defective memory when in fact they are so much more. People who never gave much thought to time travel, dimensions/multiverses, or the many world theories are dealing with the fact that the past as they remember it doesn't even remotely resemble the present and its recognized history.

I don't believe this is caused by time travel, as when the past is changed, so would be the memories of the past to reflect the new change. You cannot travel through time, change the past, and expect to see evidence that such a change ever occurred. I also believe from my research that a person can only time travel if it is possible to the point where they were born as themselves. By going past that point they could alter their own birth circumstances, so they would possibly not exist. This would then make it impossible for that person (who no longer exists) to go through time in the first place.

The theory is that if you go pass your conception, there would be different "sperm and egg events", and you would be someone else entirely. There is also the possibility that you would not have been conceived at all. People who have children would have to bear in mind that time traveling past the birth of their own children could possibly have the same effect. You could leave home and kiss your two daughters goodbye, only to return home to your three sons who would be complete strangers to you. Or the possibility exist you could go home to an empty house having had no children, by having gone past those events in time.

The Mandela Effect is not about going back in time in this dimension/ reality/universe to alter things. It is about reality shifting a group of people into another reality/dimension/ universe that – due to dissimilar events, places and things will cause individuals to have their own valid memories that will conflict with their new reality/dimension/universe.

To my understanding the ONLY evidence that this is real is the collective memories of thousands if not millions of individuals previously unconnected but now sharing the same past memories.

A theory is a hypothesis pending evidence. When evidence challenges a theory, it's time to change the theory, not time to change the evidence to fit the hypothesis. For me the massive amount of remembered information that is significantly different from this current present reality dimension is sufficient evidence to support the hypothesis that we are no longer in our original dimension/ reality. Any two realities that share a timeline close enough can stream-slip together. Anything that is the same between the two will merge. There is so far no way to say how that is determined.

Teleportation is interrelated to movement by frequencies since they are simply vibrational energies. As vibration energy at certain frequencies can affect particles causing them to separate or to change shapes, we know that movement is one of its effects. It is only a matter of "time" before various particles can exchange locations spontaneously or with assistance. I believe that many of us have teleported with assistance into this other earth; located far from our original earth/dimension/ reality.

Since everything in the universe is, at its core pure energy as in electrons, protons and neutrons it stands to reason that if what I've learned about these massive machines magnetic fields could turn out to be a lot more destructive than what is going on in the experiments themselves.

Because of this experience I can guarantee that you are going to see the world in a completely different way. There are some puzzles about the past that will now start making sense. It is rather amazing that such sophisticated people, as we judge ourselves to be, do not even know who or where we really are.

A "lie" is defined as: deliberate choice to mislead a target without giving notification of intent to do so. The important act is that a lie is an "act" that is "deliberate or intentional" and that it has the consequence of the target being "mislead."

The problem is that the longer a lie endures the more likely it is to be considered the truth unless challenged. The thought that a truth can be seen as some kind of conspiracy, marks a sad façade on the light of absolute truth.

Sadly, very few people know the real facts enough to explain them because they are not researchers themselves. As information comes down the track it becomes inaccurate and confused because most of the people passing along the information do not know the entire story, since they have not researched much information themselves, and they tend to be biased as well.

This is because your heart will tell you one thing and your intellect another and usually we get caught up believing what we are taught to believe because it's too much trouble to see things all the way through for ourselves. The verifiable information presented here may at first disturb you. It may even change the way you look at the world.

Our enemies think we are stupid, not because they are smarter, it is just that they do not understand the power of truth and justice. They only understand evil and manipulation. Like the wolves that wear sheep's clothing, they appeared to pretend to be for our good and to be harmless, longing to teach us about our world and ourselves; but they are dangerous and have continually taught a lie. However this is their function in the scheme of things, because that is what wolves do. They hunt and kill.

What I hope I have done with this book is to make you understand the necessity to search for the truth and not pretend because some people are not sharing your experiences that you are somehow mistaken.

Sadly many people have a dogmatic and dangerously arrogant attitude that all there is to know they have already learned, and they feel threatened to even hear of anything different.

There will always be human differences in perception and interpretation of events, but just as we need a common understanding of words to communicate, we need a common understanding of decency to behave decently. So be courteous in all of your dealings with others. You don't like it when someone argues with you. So don't argue with them either.

What I am condemning in this book is not the pursuit of self-interest and self-elevation, but the pursuit of self-interest and self-elevation at the expense and rights of others.

The point I am trying to make here is that whether it concerns ideas, or just life, thinking of others and being decent can make your life so much easier. You must be careful not to offend people. Keep in mind that you are probably not aware of all the knowledge in the world and neither are they. For many people these are terrifying ideas.

You can approach this topic in an intelligent manner by asking lots of questions. It's called the Socratic Method. The Socratic Method shows by means of a series of questions and answers the rational accuracy of a definition, or a point of view, or the meaning of a theory. It is based on logical analysis and will allow a person to come to the conclusion on their own without forcing the point.

If you are willing to explore evidence that contradicts the current accepted ideas we have been taught on the nature of reality and life, then this book will assist in your search.

We are living in a time of great deception, yet even that is coming to an end. However, "a little knowledge is a dangerous thing," and that is exactly what is now occurring all over the earth. So despite my extensive reading, rereading, researching and book writing I am still learning and there is so much more to discover.

Other Books By Roshan Cipriani

Rise - Be True To Yourself-Inspire Others To Live
How To Get Through Any Wall In Your Life
Train Up A Child – A Scriptural Guide To Parenting
The Art Of War For Parenting Your Teenage Child- How To Win A War You Didn't Even Know You Were In
The Key To This Life - Conscious Faith In An Unconscious World
Destiny – Past Present Future
The Seven Pillars Of Wisdom –A Sabbath Celebration Guide
Life Lessons Learned
In The Fire – Accessing Miracle Power During A Crisis
The Kingdom Lifestyle - Living By Faith And Not By Sight
God's Secret Wisdom –Principles And Secrets Of The Kingdom Of God
The Greatest Principle - The Kingdom Of God And Biblical Economics
Bricks Without Straw- Spoiling Egypt And Spoiling Babylon; The Mighty Wealth Transfer
When Failure is Not An Option
Real Faith – How To Have It And Why It Matters
The Bibles Healing Promises
I Say What They Said- Miracle Bible Prayers
The Psychology Of Stress-Dismantling The Enemy's Weapon Now
Never Quit-The Secret To Getting Through Any Wall In Your Life
His Poetry Store
SMALL BUSINESS SUCCESS- How To Write A Book Every Weekend
The Seventy Two Lunar Sabbaths- Sabbath Observance By The Phases Of The Moon
BUSINESS PLAN: Make God Your Partner –He Commanded His Blessings
PROSPERITY CONSCIOUSNESS – Living In An Abundant Universe (Personal Biblical Economics) Volume 1
Metamorphosis-Mirrors Of The Soul, Awakening To The Real You
Waiting In Goshen
How To Be Smart And Have Common Sense
None Of These Diseases –Sickness And Genocide In Second Egypt
Patience To Inherit The Promises- How To Stand By Faith Until Manifestation
The Lord Is My Shepherd, I Shall Not Want- Personal Biblical Economics
DIVORCE RECOVERY: How To Live Again
UFO COVER-UP: Biblical Evidences Uncovered-(Conspiracy) Volume 1
12 Easy Vegetarian Recipes-Healthy And Inexpensive
TRAVEL: How To Behave On An Airplane
NINJA SMOOTHIES: 21 Green Weight Loss Smoothies For The Ninja Professional Blender
Second Exodus From Second Egypt
Asset Protection And Wealth Management-Volume 1 -Trust And LLC For Legal Asset Protection
THE LOST HISTORY OF THE WORLD – Volume 1
ASSET PROTECTION 2: Wealth Management For Global Living
DISCERNMENT: The Awakening Of Real Israel
TO KNOW OURSELVES ONCE AGAIN: Your Future In Real Israel
Last Days Wisdom: FOR REAL ISRAEL
RELATIONSHIP RESCUE FROM THE BIBLE: What The Bible Says About Relationships
Living In A Fractured Multiverse-The Reality Shift Effect
Living Between The Dimensions: More Reality Change Effects
The New Dimensional Reality
Lost In The Arms Of Orion: Technology, The Mandela Effect And Parallel Dimensions

www.ingramcontent.com/pod-product-compliance
Lightning Source LLC
Chambersburg PA
CBHW080712190526
45169CB00006B/2350